Presumptive Conundrums

"You do the math. You can mix 'em and you can match 'em. There are hundreds of theories."
—Lou Jane Temple

GIVEN:
- "At reentry velocities, it only takes a small projectile to knock out a hoverstrut-supported building. So if a cylinder splits into thousands of pieces . . . well, you do the math. Or better yet, ask that woman over there to do it. The one with the puzzle cube" (Scott Westerfeld, *Extras*, 2007, p. 208).

PROVE:
Quantify the area covered by the cylinder's projectiles.

GIVEN:

- "At reentry velocities, it only takes a small projectile to knock out a hoverstrut-supported building. So if a cylinder splits into thousands of pieces ... well, you do the math. Or better yet, ask that woman over there to do it. The one with the puzzle cube" (Scott Westerfeld, *Extras*, 2007, p. 208).

PROVE:

Quantify the area covered by the cylinder's projectiles.

According to cylindrical projection, the entire world would be covered.

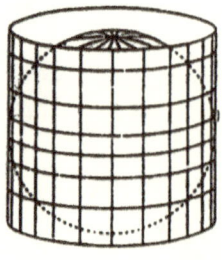 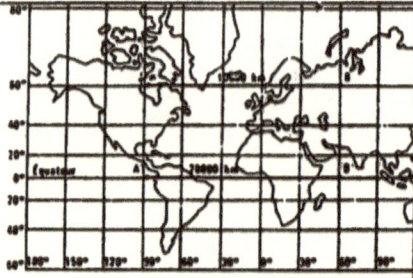

GIVEN:
- Person X was "born April second, 1908. You do the math" (Maddy Hunter, *Top O' the Mournin'*, 2003, p. 144).
- Today's date is variable.

PROVE:
Age of person X.

GIVEN:
- Person X was "born April second, 1908. You do the math" (Maddy Hunter, *Top O' the Mournin'*, 2003, p. 144).
- Today's date is variable.

PROVE:
Age of person X.

April 2nd, 1908

$$8 + 4 - 9 - 2 - 1 = 0$$

Age at birth: zero

GIVEN:

- "Worms by themselves only travel a few meters a year. You do the math—it would take them about a hundred years to travel a quarter-mile" (Amy Stewart, *The Earth Moved*, 2005, p. 108).

PROVE:
True or False.

GIVEN:
- "Worms by themselves only travel a few meters a year. You do the math—it would take them about a hundred years to travel a quarter-mile" (Amy Stewart, *The Earth Moved*, 2005, p. 108).

PROVE:
True or False.

False. They may travel a greater distance in less time via a Lorentzian traversible wormhole.

The metric solution is:

$$ds^2 = -c^2 dt^2 + dl^2 + (k^2 + l^2)(d\theta^2 + \sin^2\theta \, d\phi^2)$$

GIVEN:
- "All I can say is . . . The Spot and Valentine's Day. You do the math" (Brian Peterson, *Move Over, Girl*, 1998, p. 34).

PROVE:
Derive the equation.

GIVEN:
- "All I can say is . . . The Spot and Valentine's Day. You do the math" (Brian Peterson, *Move Over, Girl*, 1998, p. 34).

PROVE:
Derive the equation.

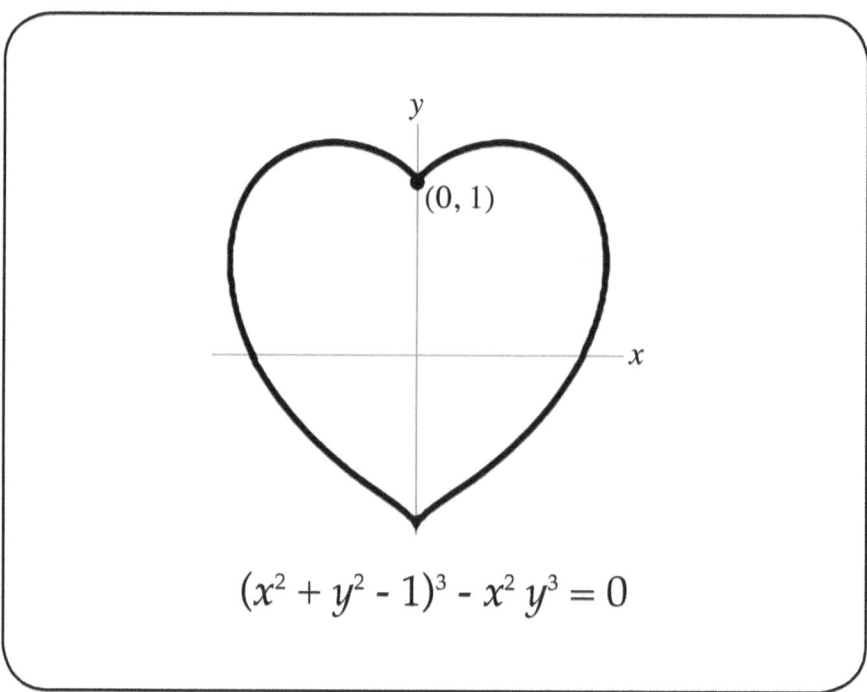

$$(x^2 + y^2 - 1)^3 - x^2 y^3 = 0$$

GIVEN:

- "'Two Ears, One Mouth,' the note read. He was mystified. Was it an insult? Was it a warning? Was it in code? Did it contain a secret message?" (Chenjerai Hove, *Palaver Finish*, 2002, p. 77).
- "Two ears, one mouth—you do the math" (Chip Bell, *Managing Knock Your Socks Off Service*, 1992, p. 36).

PROVE:
Represent mathematically.

GIVEN:

- "'Two Ears, One Mouth,' the note read. He was mystified. Was it an insult? Was it a warning? Was it in code? Did it contain a secret message?" (Chenjerai Hove, *Palaver Finish*, 2002, p. 77).
- "Two ears, one mouth—you do the math" (Chip Bell, *Managing Knock Your Socks Off Service*, 1992, p. 36).

PROVE:

Represent mathematically.

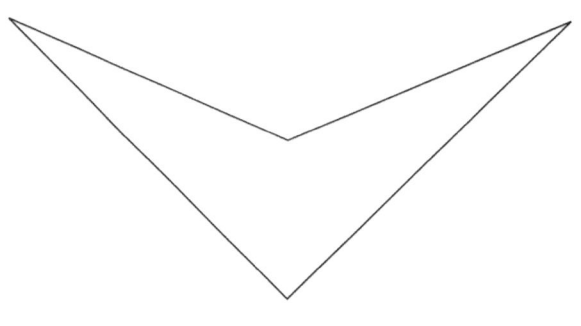

An Anthropomorphic Polygon contains precisely two "ears" and one "mouth."

GIVEN:
- "Men in tight leotards, divinely muscled women who sweat in public, plus music and rapt onlookers? You do the math" (Kathie Bergquist, *A Field Guide to Gay & Lesbian Chicago*, 2006, p. 180).

PROVE:
Calculate Orchestra, Mezzanine, and Balcony ticket prices.

GIVEN:
- "Men in tight leotards, divinely muscled women who sweat
  in public, plus music and rapt onlookers? You do the math"
  (Kathie Bergquist, *A Field Guide to Gay & Lesbian Chicago*,
  2006, p. 180).

PROVE:
Calculate Orchestra, Mezzanine, and Balcony ticket prices.

Indeterminate. Prices vary according
to season, location, and incentive
programs, and are subject to change
without prior notice.

GIVEN:

- Person X "was a scientist. And he was killed. You do the math" (Boris Starling, *Visibility*, 2006, p. 158).
- All survivors of Person X's wound sought modern medical attention.

PROVE:
What type of scientist is Person X?

GIVEN:
- Person X "was a scientist. And he was killed. You do the math" (Boris Starling, *Visibility*, 2006, p. 158).
- All survivors of Person X's wound sought modern medical attention.

PROVE:
What type of scientist is Person X?

X is a Christian Scientist.

Additional premise:

☠ No Christian Scientists seek modern medical treatment.

18

GIVEN:
- "I saw two individuals leave with Matthew. I didn't see their faces, but I saw the back of their heads. At the same time, McKinney and Henderson were no longer around. You do the math" (Moises Kaufman, *The Laramie Project*, 2001, p. 41).

PROVE:
Represent algebraically. What must be true about Matthew, McKinney, and Henderson?

GIVEN:
- "I saw two individuals leave with Matthew. I didn't see their faces, but I saw the back of their heads. At the same time, McKinney and Henderson were no longer around. You do the math" (Moises Kaufman, *The Laramie Project*, 2001, p. 41).

PROVE:
Represent algebraically. What must be true about Matthew, McKinney, and Henderson?

Let x = a known person

Let y = an unknown person

It is claimed that $1x + 2y = 3x$

This is true only if $y = x$.

Matthew, McKinney, and Henderson must be conjoined triplets.

GIVEN:

- "She was a sophisticated woman of a certain age . . . living alone in Greenwich Village. I mean, you do the math" (Lawrence Block, *All the Flowers are Dying*, 2005, p. 132).

PROVE:
Was she kinky?

GIVEN:

• "She was a sophisticated woman of a certain age ... living alone in Greenwich Village. I mean, you do the math" (Lawrence Block, *All the Flowers are Dying*, 2005, p. 132).

PROVE:

Was she kinky?

Let X = the neighborhood's haphazard layout, with streets curving at odd angles.

Let Y = "a certain age," implying osteoporosis, a bending of the spine.

Let Z = "sophisticated," meaning complicated, not straightforward.

∴ She was kinky.

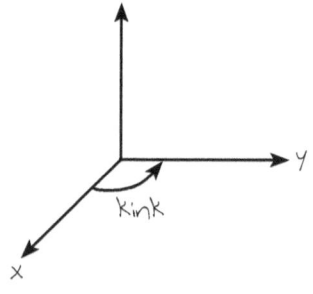

GIVEN:
- "Five times a day times seven days a week plus two small hands clenched together in fear and ignorance / equals a lifetime of trying to make halves a whole" (Alix Olson, "You Do the Math," *Word Warriors*, 2007, p. 93).

PROVE:
The author's success rate.

GIVEN:

- "Five times a day times seven days a week plus two small hands clenched together in fear and ignorance / equals a lifetime of trying to make halves a whole" (Alix Olson, "You Do the Math," *Word Warriors*, 2007, p. 93).

PROVE:
The author's success rate.

Zero percent success.

It is not possible for any size hands to divide any number of years into a whole number of weeks, let alone an even number.

GIVEN:

- "My show is heard by millions of listeners. Each of them would like a few moments of my time. A few moments multiplied by a few million listeners ... you do the math" (Alton Gansky, *The Prodigy*, 2001, p. 105).

PROVE:

Represent mathematically.

GIVEN:

- "My show is heard by millions of listeners. Each of them would like a few moments of my time. A few moments multiplied by a few million listeners . . . you do the math" (Alton Gansky, *The Prodigy*, 2001, p. 105).

PROVE:

Represent mathematically.

One moment equals 65 instants, according to the Buddhist *Abhidharma* scriptures.

The duration of an instant is near-zero.

White noise is a radio signal of infinite power at near-zero time shift.

$$R_{ww}(t_1, t_2) = E\{w(t_1)w(t_2)\} = (N_0/2)\delta(t_1 - t_2)$$

GIVEN:

• "There are nine million of us and only one of you. Do the math, Raymond. It's not hard to figure the odds" (Allan Folsom, *The Exile*, 2004, p. 100).

PROVE:

Reports of Raymond's sudden death at the hands of nine million people are "fishy."

GIVEN:

• "There are nine million of us and only one of you. Do the math, Raymond. It's not hard to figure the odds" (Allan Folsom, *The Exile*, 2004, p. 100).

PROVE:

Reports of Raymond's sudden death at the hands of nine million people are "fishy."

The probabilities of Raymond's sudden death follow a "Poisson" Distribution:

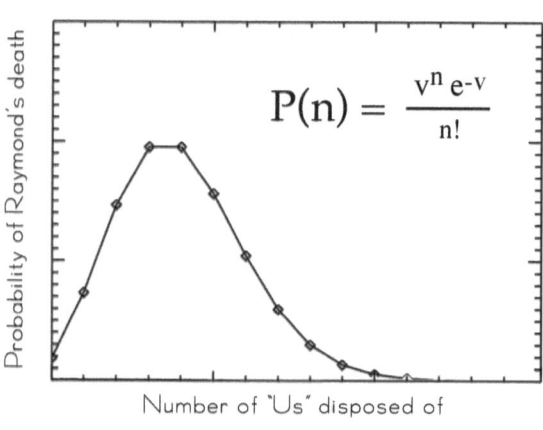

$$P(n) = \frac{v^n e^{-v}}{n!}$$

Probability of Raymond's death

Number of "Us" disposed of

GIVEN:

- "I had lost my job at the base PX and I had lost my gag reflex. You do the math" (Stephen Trask and John Cameron Mitchell, *Hedwig and the Angry Inch*, 2003, p. 25).

PROVE:

How is a lost reflex directly proportional to unemployability?

Check Your Work

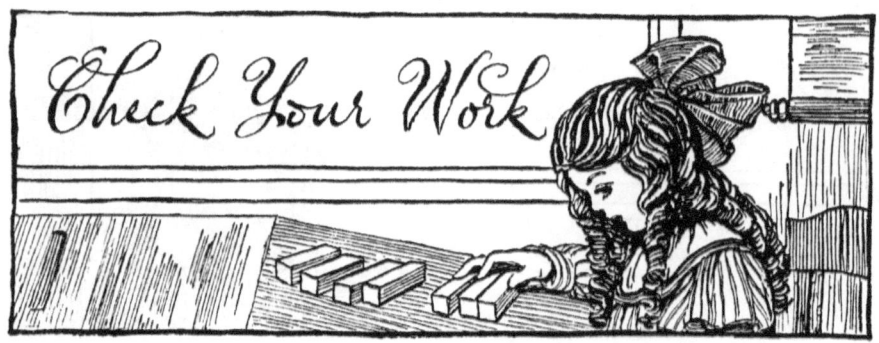

GIVEN:

- "I had lost my job at the base PX and I had lost my gag reflex. You do the math" (Stephen Trask and John Cameron Mitchell, *Hedwig and the Angry Inch*, 2003, p. 25).

PROVE:

How is a lost reflex directly proportional to unemployability?

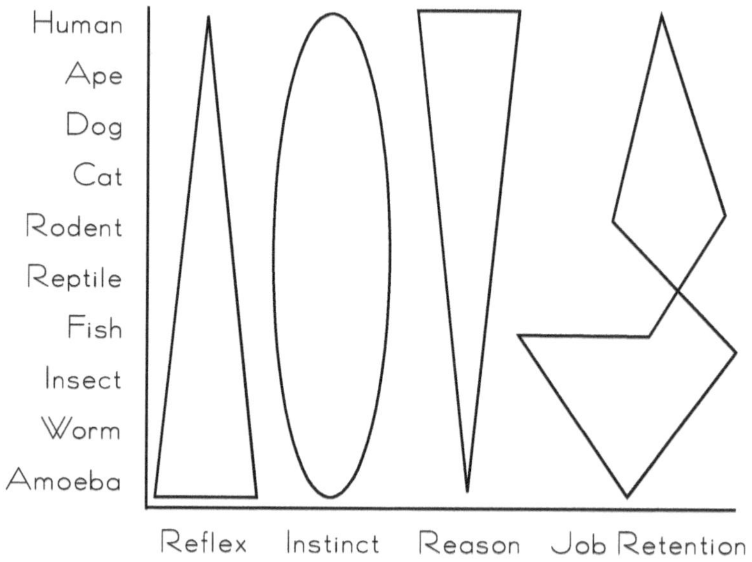
Human
Ape
Dog
Cat
Rodent
Reptile
Fish
Insect
Worm
Amoeba
Reflex
Instinct
Reason
Job Retention

GIVEN:

- "Five kids, ages seven, five, three, two, and six months. A husband who works full time at NASA as a rocket scientist. A troubled, homeschooling mom. You do the math. It adds up to insanity" (Janet Velez-Mitchell, *Secrets Can Be Murder*, 2007, p. 54).

PROVE:
True or false.

GIVEN:

• "Five kids, ages seven, five, three, two, and six months. A husband who works full time at NASA as a rocket scientist. A troubled, homeschooling mom. You do the math. It adds up to insanity" (Janet Velez-Mitchell, *Secrets Can Be Murder*, 2007, p. 54).

PROVE:
True or false.

> 364 weeks (age of 7-year-old)
> +260 weeks (age of 5-year-old)
> +156 weeks (age of 3-year-old)
> +104 weeks (age of 2-year-old)
> +26 weeks (age of 6-month-old)
> +.238... week (40-hours full-time job)
> +3.428... (24/7 teaching)
>
> = 913.$\overline{6}$ (irrational number)
>
> True.

GIVEN:
- "Ma's Buddhist. Dad's atheist. A Methodist luncheon. You do the math" (Kip Fulbeck, *Paper Bullets*, 2001, p. 11).

PROVE:
What time is food served?

GIVEN:

- "Ma's Buddhist. Dad's atheist. A Methodist luncheon. You do the math" (Kip Fulbeck, *Paper Bullets*, 2001, p. 11).

PROVE:

What time is food served?

> 8 (Buddhist noble eightfold path)
> + 3 (Methodist trinity)
> - {} (Atheism)
>
> = 11 (lunch hour)

GIVEN:
- "So you do the math: sugar = fat" (Rory Freedman, *Skinny Bitch*, 2007, p. 18).

PROVE:
Represent mathematically.

GIVEN:
- "So you do the math: sugar = fat" (Rory Freedman, *Skinny Bitch*, 2007, p. 18).

PROVE:
Represent mathematically.

Let sugar = fructose

Let fat = adipose

According to alphabetic numerology:

ADIPOSE → 1+4+9+7+6+1+5=33 → 3+3=6

FRUCTOSE → 6+9+3+3+2+6+1+3=33 → 3+3=6

Q.E.D.

GIVEN:
- "Now you do the math: If there are two women fighting over every guy in lockdown, that means four are fighting over every bus driver and eight are fighting over every college professor, and God knows how may are fighting over an investment banker" (Linda Villarosa, *Passing for Black*, 2008, p. 17).

PROVE:
Most men are untouchables.

GIVEN:

• "Now you do the math: If there are two women fighting over every guy in lockdown, that means four are fighting over every bus driver and eight are fighting over every college professor, and God knows how may are fighting over an investment banker" (Linda Villarosa, *Passing for Black*, 2008, p. 17).

PROVE:

Most men are untouchables.

Let W = {all women} and M = {all men}.

Divide W into subsets of 2+4+8+N women and M into corresponding sets of 4 men (e.g. prisoner, bus driver, professor, and banker).

Suppose |W| = |M| (i.e. there are as many men as women).

Then there are (14+N)/4 women fighting over any desirable man.

Conversely, the ratio of undesirable men to desirable men is (14+N-4)/(14+N), which is greater than 50% for any N, not even taking into consideration women's compounding interest in the bankers.

- "He says angrily, 'Nothing about any of this is fair. Do you want to know what's fair? *Nothing.* Not one fucking thing. I'm not fair, and Nan isn't fair, and Christopher, for damn sure, isn't fair. You're not all that fair either, are you? Do the math" (Stacey D'Erasmo, *A Seahorse Year*, 2005, p. 269).

PROVE:
Show that the speaker is incorrect.

GIVEN:

- "He says angrily, 'Nothing about any of this is fair. Do you want to know what's fair? *Nothing.* Not one fucking thing. I'm not fair, and Nan isn't fair, and Christopher, for damn sure, isn't fair. You're not all that fair either, are you? Do the math." (Stacey D'Erasmo, *A Seahorse Year*, 2005, p. 269).

PROVE:

Show that the speaker is incorrect.

Suppose X is fair $\Leftrightarrow$ X$\neq$0 for some X.

Then Nan = 0, Christopher = 0, speaker = 0, and you = 0.

Let F = {Nan, Christopher, speaker, you}

|F| = 4 $\neq$ 0 $\Rightarrow$ F is fair.

"Nothing is good or fair alone."
—Ralph Waldo Emerson

GIVEN:

- "And you had better make sure that you do the math right. But that will only be as good as the winds allow it to be. And winds can be capricious" (James Doyle, *Flying Through Time*, 2003, p. 80).

PROVE:
17 < five knots < 21.

GIVEN:

• "And you had better make sure that you do the math right. But that will only be as good as the winds allow it to be. And winds can be capricious" (James Doyle, *Flying Through Time*, 2003, p. 80).

PROVE:
17 < five knots < 21.

The Beaufort Wind Scale (in knots):

0 = <1
1 = 1-3
2 = 4-6
3 = 7-10
4 = 11-16
5 = 17-21
6 = 22-27
7 = 28-33
8 = 34-40
9 = 41-47
10 = 48-55
11 = 56-63
12 = 64+

"One drifts as the winds allow."
—James Whitcomb Riley

GIVEN:

- "I'm in a fucking psychiatric ward ... I'm wearing a fucking wristband with my name on it, and I'm taking antipsychotics ... so you do the math" (Lynn Marie Smith, *Rolling Away*, 2005, p. 151).

PROVE:
The odds.

GIVEN:

- "I'm in a fucking psychiatric ward ... I'm wearing a fucking wristband with my name on it, and I'm taking antipsychotics ... so you do the math" (Lynn Marie Smith, *Rolling Away*, 2005, p. 151).

PROVE:
The odds.

28%.

The average rate of success of insanity pleas is 28%, according to *The Handbook of Forensic Psychology.*

GIVEN:

- "I've seen an epidural turn a snarling, thrashing alligator into a kitten in the span of five minutes. You do the math" (Scott Mactavish, *The New Dad's Survival Guide*, 2005, p. 7).

PROVE:
Such an epidural cannot exist.

GIVEN:

- "I've seen an epidural turn a snarling, thrashing alligator into a kitten in the span of five minutes. You do the math" (Scott Mactavish, *The New Dad's Survival Guide*, 2005, p. 7).

PROVE:

He should have said "turn a kitten into an alligator."

The metabolic rate of a kitten (250 kcal/kg) is 243 times greater than that of an alligator (1.03 kcal/kg).

GIVEN:
- "It's a considerable amount of money. What, six hundred guys times nine hundred bucks? You do the math" (Elda Minger, "Mr. Speedy," *Fantasy*, 2005, p. 221).

PROVE:
The ramification.

GIVEN:
- "It's a considerable amount of money. What, six hundred guys times nine hundred bucks? You do the math" (Elda Minger, "Mr. Speedy," *Fantasy*, 2005, p. 221).

PROVE:
The ramification.

27,000 lap dances.

The average lap dance costs $20 per three minutes.

600
x $900
÷ $20
= 27,000 lap dances (1,350 hours)

GIVEN:

• "I am writing an equation. / Using the universal language of numbers to describe ten thousand ways / that something can mean everything" (Meliza Bañales, "Do the Math," *Word Warriors*, 2007, p. 94).

PROVE:

Represent the equation.

GIVEN:

• "I am writing an equation. / Using the universal language of numbers to describe ten thousand ways / that something can mean everything" (Meliza Bañales, "Do the Math," *Word Warriors*, 2007, p. 94).

PROVE:

Represent such an equation.

(The original Greek meaning of "myriad" was "ten thousand.")

Consider the N Harmonic series given by

$$\sum_{k=1}^{\infty} \frac{1}{kN}$$

Even though $\frac{1}{kN}$ approaches zero as k approaches infinity (each term being just a 'little something') the sum of myriad somethings is infinite.

GIVEN:
- "Once you do the math, it becomes clear that you're better off taking your chances on the wolf than signing up for a guaranteed skinning by your protector" (Alexander Elder, *Come Into My Trading Room*, 2002, p. 27).

PROVE:
True or false.

GIVEN:
- "Once you do the math, it becomes clear that you're better off taking your chances on the wolf than signing up for a guaranteed skinning by your protector" (Alexander Elder, *Come Into My Trading Room*, 2002, p. 27).

PROVE:
True or false.

According to Wolf's theory, "you are free only when you are doing the *right* thing for the *right* reasons" (Robert Kane, *The Oxford Handbook of Free Will*, 2005, p. 21).

You Do The Math

GIVEN:

- "You do the math. Murder plus mayor equals ratings" (Thomas Cavanagh, *Prodigal Son*, 2008, p. 95).

PROVE:
Represent mathematically.

GIVEN:

- "You do the math. Murder plus mayor equals ratings" (Thomas Cavanagh, *Prodigal Son*, 2008, p. 95).

PROVE:
Represent mathematically.

Murder Formula[1]
$$hr = (100,000 / N_p) \, N_h$$
where hr equals homicide rate, N_p equals population, and N_h equals number of homicides

+ Mayoral Corruption Formula[2]
$$C = M + D - A$$
Corruption equals Monopoly power plus Discretion by officials minus Accountability

= Ratings Calculation
One rating point is worth 1,000,000 households

[1] Michael A. Bellesiles, *Lethal Imagination*, 1999, p. 213
[2] Maria Gonzalez De Asis, *Transparency and Accountability in Water and Sanitation*, 2009, p. 16

GIVEN:
- "You do the math and see which side you come out on. Live poor. Die rich. You do the math" (Damella Ford, *Naked Love*, 2007, p. 185).

PROVE:
Plot on a graph.

GIVEN:
- "You do the math and see which side you come out on. Live poor. Die rich. You do the math" (Damella Ford, *Naked Love*, 2007, p. 185).

PROVE:
Plot on a graph.

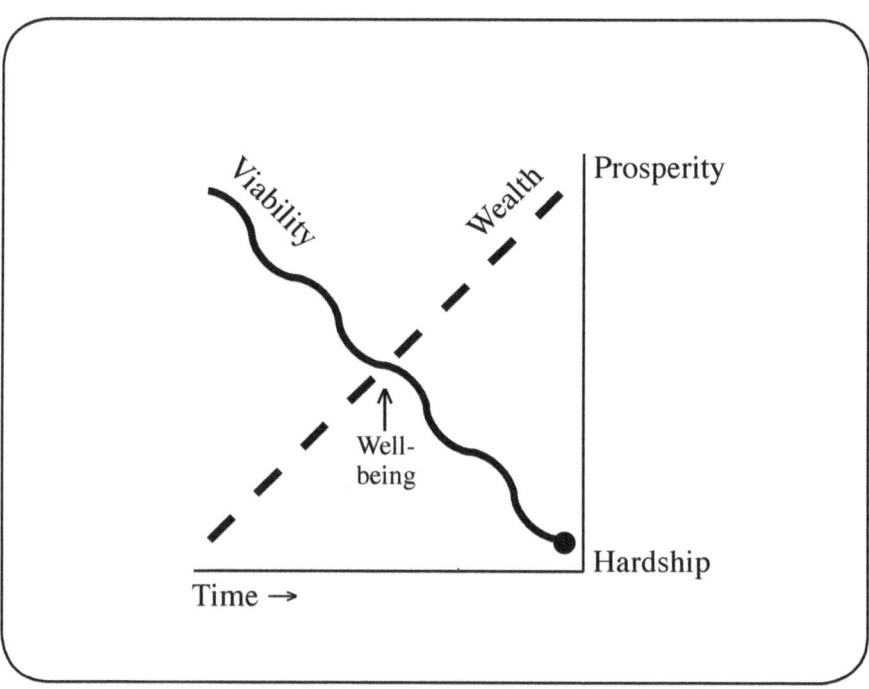

GIVEN:

- "It's about numbers. Compared to revenue, the expenses are nothing. You do the math" (Sigmund Brouwer, *Crown of Thorns*, 2002, p. 164).

PROVE:

Express mathematically using Generally Accepted Accounting Principles.

Check Your Work

GIVEN:

• "It's about numbers. Compared to revenue, the expenses are nothing. You do the math" (Sigmund Brouwer, *Crown of Thorns*, 2002, p. 164).

PROVE:

Express mathematically using Generally Accepted Accounting Principles.

Let O = owner's equity
Let L = liabilities
Let A = assets
Let R = revenue
Let E = expenses

The journal balancing equation is

$$O + (L - A) + (R - E) = \text{zero}$$

GIVEN:
- "Mom, dead people are talking to you. Do the math"
  ("Gingerbread," *Buffy the Vampire Slayer*, Nov. 9, 1998, p. 47).

PROVE:
Represent in American Standard Code for Information Interchange.

GIVEN:
• "Mom, dead people are talking to you. Do the math"
  ("Gingerbread," *Buffy the Vampire Slayer*, Nov. 9, 1998, p. 47).

PROVE:
Represent in American Standard Code for Information
Interchange.

```
--------------------------------
|                                |
|      YES          NO           |
|                                |
|     ABCDEFGHIJKLM              |
|     NOPQRSTUVWXYZ              |
|                                |
|                                |
|     1234567890                 |
|                                |
|                                |
|      x GOOD BYE x              |
|                                |
--------------------------------
       ASCII OUIJA BOARD™
```

GIVEN:
- "Three doors or one window. You do the math" (Lisa Lutz, *The Spellman Files*, 2007, p. 192).

PROVE:
The differential.

GIVEN:

- "Three doors or one window. You do the math" (Lisa Lutz, *The Spellman Files*, 2007, p. 192).

PROVE:
The differential.

Zero differential.

The arc of the swing of three doors (including one two-way door) equals a semicircle window (π).

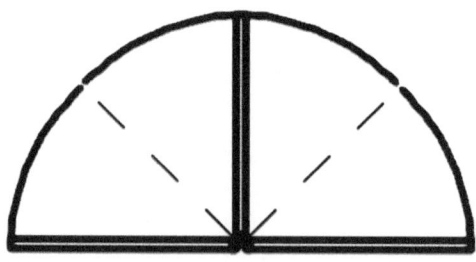

- "You're not organizing your finances here. You are looking for someone to spend the rest of your life with. You do the math" (Brenda Della Casa, *Cinderella Was a Liar*, 2006, p. 32).

PROVE:
Represent graphically.

GIVEN:

• "You're not organizing your finances here. You are looking for someone to spend the rest of your life with. You do the math" (Brenda Della Casa, *Cinderella Was a Liar*, 2006, p. 32).

PROVE:
Represent graphically.

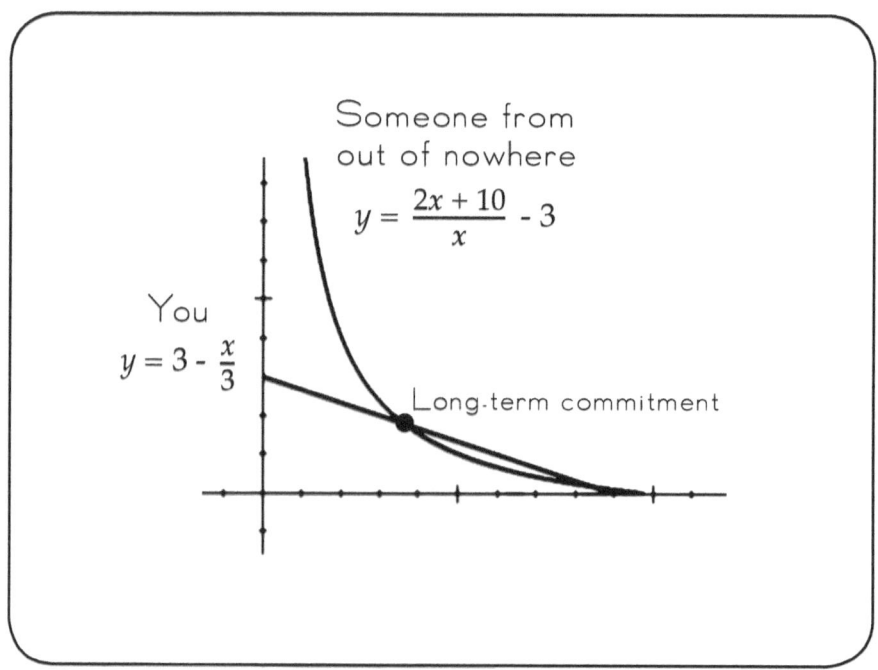

Someone from out of nowhere

$$y = \frac{2x + 10}{x} - 3$$

You

$$y = 3 - \frac{x}{3}$$

Long-term commitment

GIVEN:

- "Fifteen thousand of them, one of you. Do the math, huh?"
  (Robert Ludlum, *The Bancroft Strategy*, 2007, p. 535).

PROVE:

The odds ratio.

GIVEN:
- "Fifteen thousand of them, one of you. Do the math, huh?" (Robert Ludlum, *The Bancroft Strategy*, 2007, p. 535).

PROVE:
The odds ratio.

$$\frac{15,000}{\frac{1}{15,000}} = 15,000 \times 15,000 = 225,000,000$$

GIVEN:
- "Twice as long as the pair of you, three times as long as a whatever you want to call it, what do you [sic], doesn't matter, three. You do the math" (Ramsey Campbell, *The Overnight*, 2006, p. 159).

PROVE:
True or false.

GIVEN:

• "Twice as long as the pair of you, three times as long as a whatever you want to call it, what do you [sic], doesn't matter, three. You do the math" (Ramsey Campbell, *The Overnight*, 2006, p. 159).

PROVE:
True or false.

False.

2(2y) [twice the pair of you]
= 3x   [three times whatchamacallit]

x = 3
y = $\frac{3}{4}$x

⇒ y = 2.25

GIVEN:

- "She left this morning with a blond wig, and unfortunate panty lines. . . . She comes back with smeared mascara, whisker burn, no wig, and no panty lines. . . . You do the math, gentlemen" (Shannon McKenna, *Edge of Midnight*, 2007, p. 303).

PROVE:
Support the claim that a liaison took place.

GIVEN:

- "She left this morning with a blond wig, and unfortunate panty lines. . . . She comes back with smeared mascara, whisker burn, no wig, and no panty lines. . . . You do the math, gentlemen" (Shannon McKenna, *Edge of Midnight*, 2007, p. 303).

PROVE:

Support the claim that a liaison took place.

Truth Table	Morning	Night
Blond wig	T (1)	F (-1)
Smeared mascara	F (-1)	T (1)
Panty lines	T (1)	F (-1)
Whisker burn	F (-1)	T (1)
	0	0

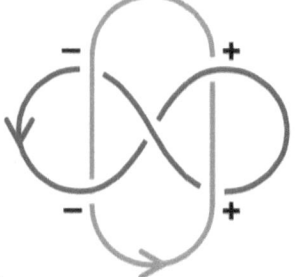

By knot theory, the net linking number of zero implies a Whitehead entanglement.

GIVEN:
- "If you do the math, you can probably guess that any news articles about her aren't going to make her life better" (Susan May Warren, *Escape to Morning*, 2005, p. 99).

PROVE:
True or false.

- "If you do the math, you can probably guess that any news articles about her aren't going to make her life better" (Susan May Warren, *Escape to Morning*, 2005, p. 99).

PROVE:
True or false.

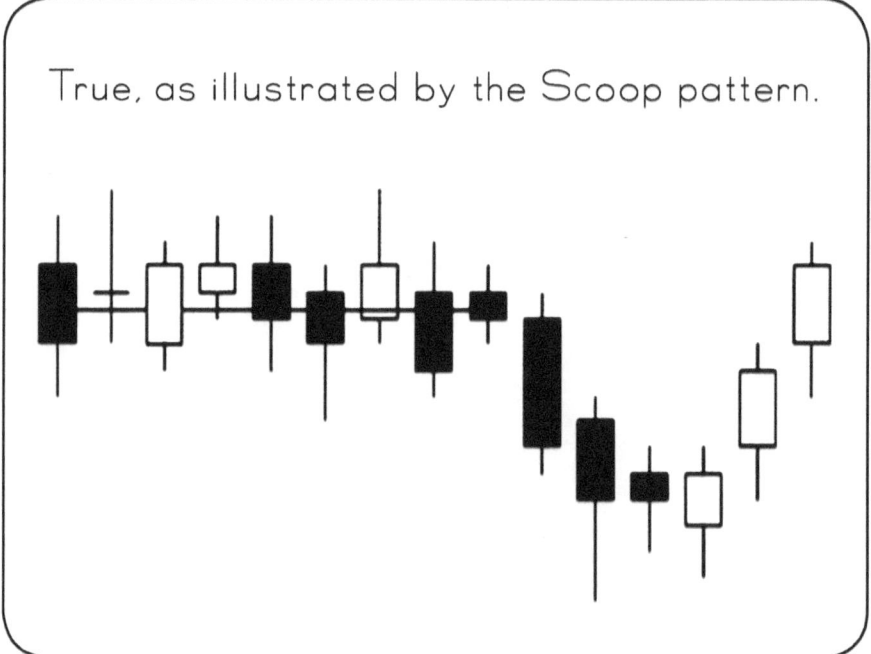

True, as illustrated by the Scoop pattern.

GIVEN:

- "Over the ten years I went to this hairdresser I brought her twelve customers—all of whom visited her at least once a month. (You do the math.)" (Chris Clarke-Epstein, *78 Important Questions Every Leader Should Ask and Answer*, 2002, p. 39).

PROVE:

The possible hairdresser appointment books are uncountable.

GIVEN:

- "Over the ten years I went to this hairdresser I brought her twelve customers—all of whom visited her at least once a month. (You do the math.)" (Chris Clarke-Epstein, *78 Important Questions Every Leader Should Ask and Answer*, 2002, p. 39).

PROVE:

The possible hairdresser appointment books are uncountable.

The order of monthly appointments made by 12 people over 10 years is a braid group consisting of 10 x 12 x 12! items.

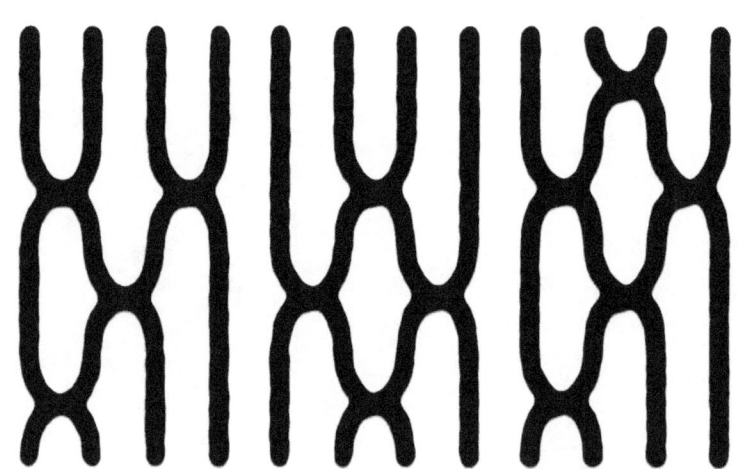

GIVEN:
- "Six years versus over 150 years. You do the math. You can't deliver results without a great culture, but a great culture without results has no value" (Randy Pennington, *Results Rule!*, 2006, p. 10).

PROVE:
Represent on a pie chart.

GIVEN:

• "Six years versus over 150 years. You do the math. You can't deliver results without a great culture, but a great culture without results has no value" (Randy Pennington, *Results Rule!*, 2006, p. 10).

PROVE:
Represent on a pie chart.

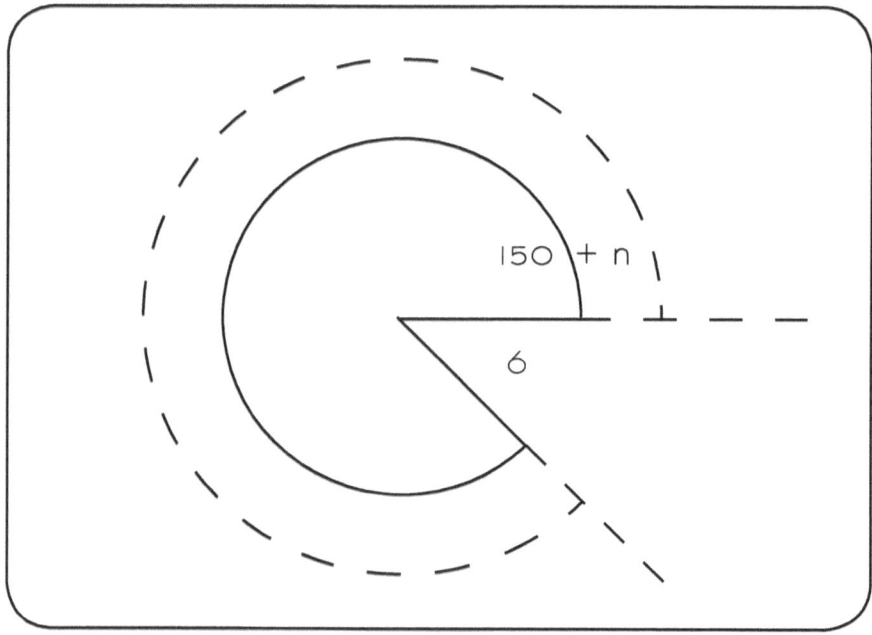

GIVEN:
- "My brother's a single man, and he's bringing a woman home. ... You do the math" (Gail McFarland, *Dream Runner*, 2008, p. 212).

PROVE:
Represent on a graph.

Check Your Work

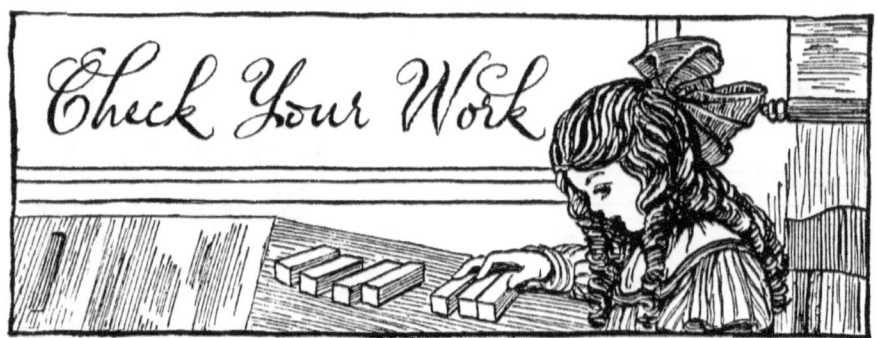

GIVEN:

- "My brother's a single man, and he's bringing a woman home. . . . You do the math" (Gail McFarland, *Dream Runner*, 2008, p. 212).

PROVE:

Represent on a graph.

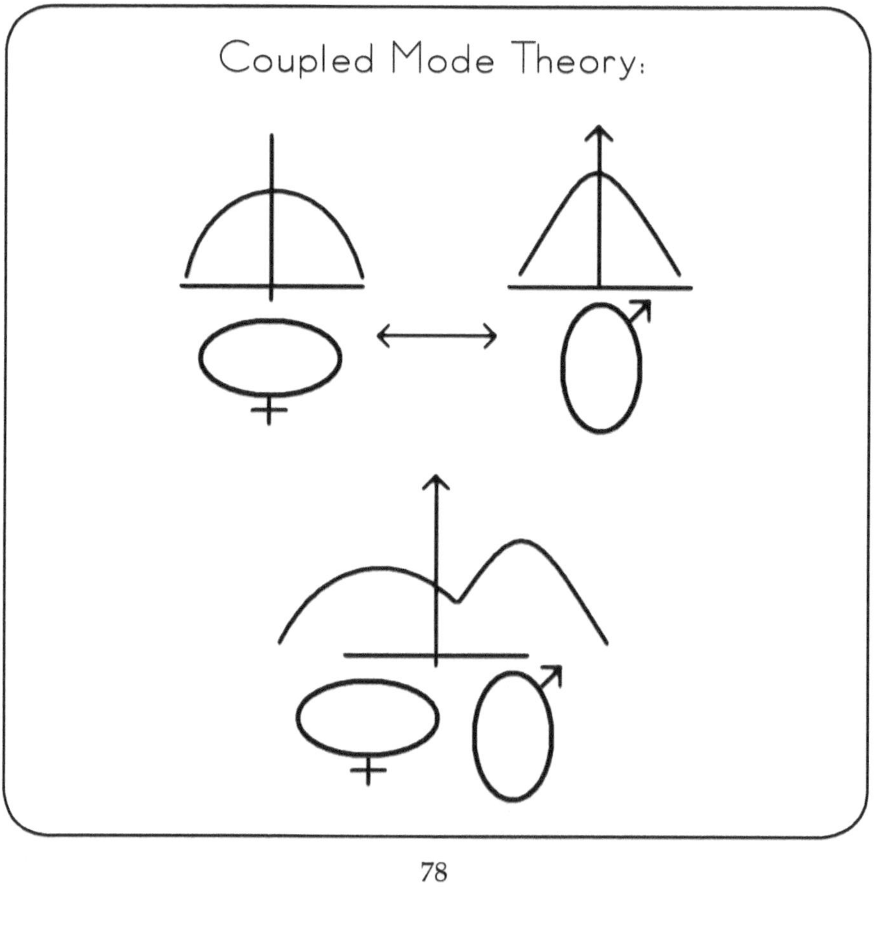
Coupled Mode Theory:

GIVEN:
- "Two martinis or a pan of lemon bars. You do the math" (Isis Crawford, *A Catered Murder*, 2004, p. 184).

PROVE:
Are they equal?

GIVEN:
- "Two martinis or a pan of lemon bars. You do the math" (Isis Crawford, *A Catered Murder*, 2004, p. 184).

PROVE:
Are they equal?

Yes, 2 = 13.

"Two equals thirteen" (a baker's dozen) is relatively closer to the truth than "two equals forty-one," to paraphrase Vladimir Ivanovich Savchenko.

GIVEN:

- "What kind of man goes after a woman who's married, a woman who's an obvious drunk, and who has a little girl? You do the math" (Jodi Picoult, *Vanishing Acts*, 2005, p. 386).

PROVE:
Represent mathematically.

GIVEN:

• "What kind of man goes after a woman who's married, a woman who's an obvious drunk, and who has a little girl? You do the math" (Jodi Picoult, *Vanishing Acts*, 2005, p. 386).

PROVE:
Represent mathematically.

Let M represent a man,
 W^2 a married woman,
 DG the state of being drunk with a daughter.

$M + (W^2 + DG) = 2$ [a couple]

$\Rightarrow M = 1 = W^2 + DG$

$\Rightarrow 1M \Leftrightarrow W^2 + DG$

∴ The one man who goes after $(W^2 + DG)$ is the only type of man satisfying the relationship condition.

GIVEN:
- "You said I didn't act like I was married. You do the math"
  (Alisa Valdes-Rodriquez, *Dirty Girls on Top*, 2008, p. 243).

PROVE:
Are they still a couple?

GIVEN:
• "You said I didn't act like I was married. You do the math"
(Alisa Valdes-Rodriquez, *Dirty Girls on Top*, 2008, p. 243).

PROVE:
Are they still a couple?

According to the Force/Separation Curve, when the forces of attraction diminish, a repelling force leads to separation.

GIVEN:
- "Every man thinks he's an exception to the statistics. . . . You do the math" (Martha I. Finney, *The Truth about Getting the Best from People*, 2008, p. 30).

PROVE:
Represent mathematically.

GIVEN:

- "Every man thinks he's an exception to the statistics. . . . You do the math" (Martha I. Finney, *The Truth about Getting the Best from People*, 2008, p. 30).

PROVE:
Represent mathematically.

"We are all like snowflakes."
—Lewis Black

$$d = \frac{\ln 4}{\ln 3}$$

△　∧
base　motif

GIVEN:
- "Weekend lunch menu's good till four, and I don't come in until two. You do the math" (Alafair Burke, *Missing Justice*, 2004, p. 339).

PROVE:
What was the tab?

GIVEN:
- "Weekend lunch menu's good till four, and I don't come in until two. You do the math" (Alafair Burke, *Missing Justice*, 2004, p. 339).

PROVE:
What was the tab?

In accordance with cosmologist Alan Guth's "Free Lunch Theory" of the universe, his meal was on the house.

GIVEN:
- "I like these girls. I don't like you. Do the math. Consider yourself subtracted" (Rachel Caine, *The Dead Girls' Dance*, 2007, p. 183).

PROVE:
The speaker likes hourglass figures.

GIVEN:

- "I like these girls. I don't like you. Do the math. Consider yourself subtracted" (Rachel Caine, *The Dead Girls' Dance*, 2007, p. 183).

PROVE:

The speaker likes hourglass figures.

Per the XY sex-determination system,
let X^2 = the girls the speaker likes
and XY = the boy the speaker doesn't like.

Then we are given
$X^2 - XY = 1$
which is a
hyperbola.

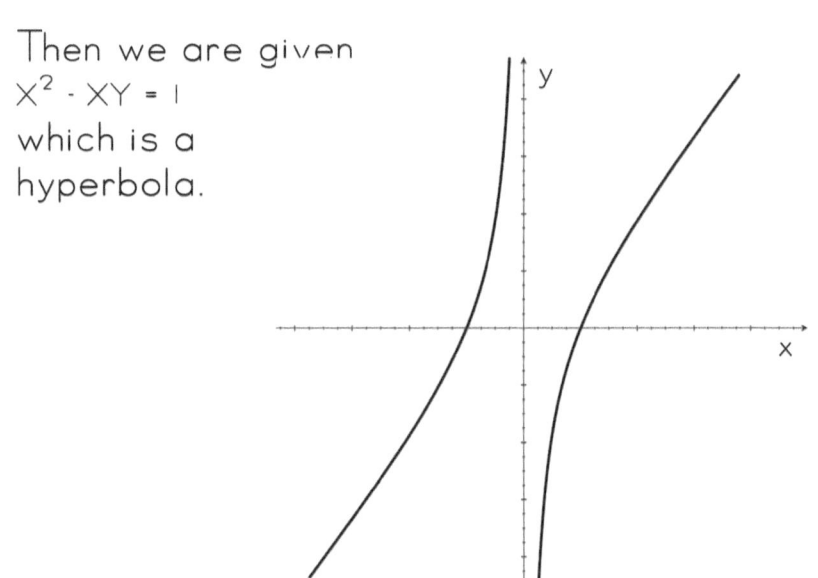

GIVEN:
- "Before jumping to conclusions, you should do the math" (James Flynn, *Roadmap to 6th Grade Math*, 2002, p. 190).

PROVE:
Represent mathematically.

GIVEN:

- "Before jumping to conclusions, you should do the math" (James Flynn, *Roadmap to 6th Grade Math*, 2002, p. 190).

PROVE:

Represent mathematically.

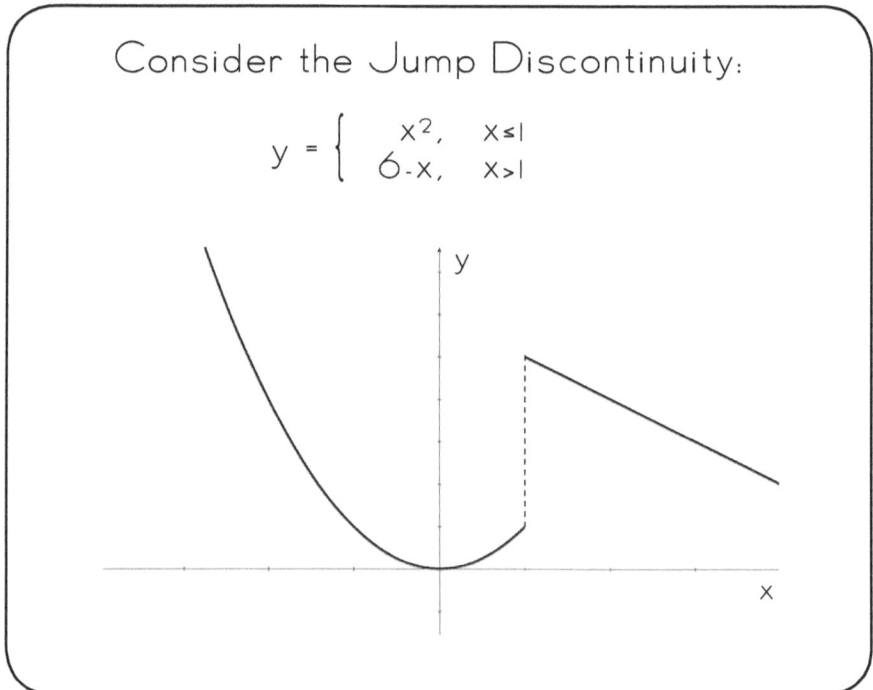

Consider the Jump Discontinuity:

$$y = \begin{cases} x^2, & x \leq 1 \\ 6\text{-}x, & x > 1 \end{cases}$$

GIVEN:

- "Do the math. With an interception velocity of 80 kilometers per second, a single grain of rice can produce a catastrophic result. Now multiply that by a hundred thousand. Or a million—" (David Gerrold, *Leaping to the Stars*, 2002, p. 109).

PROVE:
Represent mathematically.

GIVEN:

- "Do the math. With an interception velocity of 80 kilometers per second, a single grain of rice can produce a catastrophic result. Now multiply that by a hundred thousand. Or a million—" (David Gerrold, *Leaping to the Stars*, 2002, p. 109).

PROVE:
Represent mathematically.

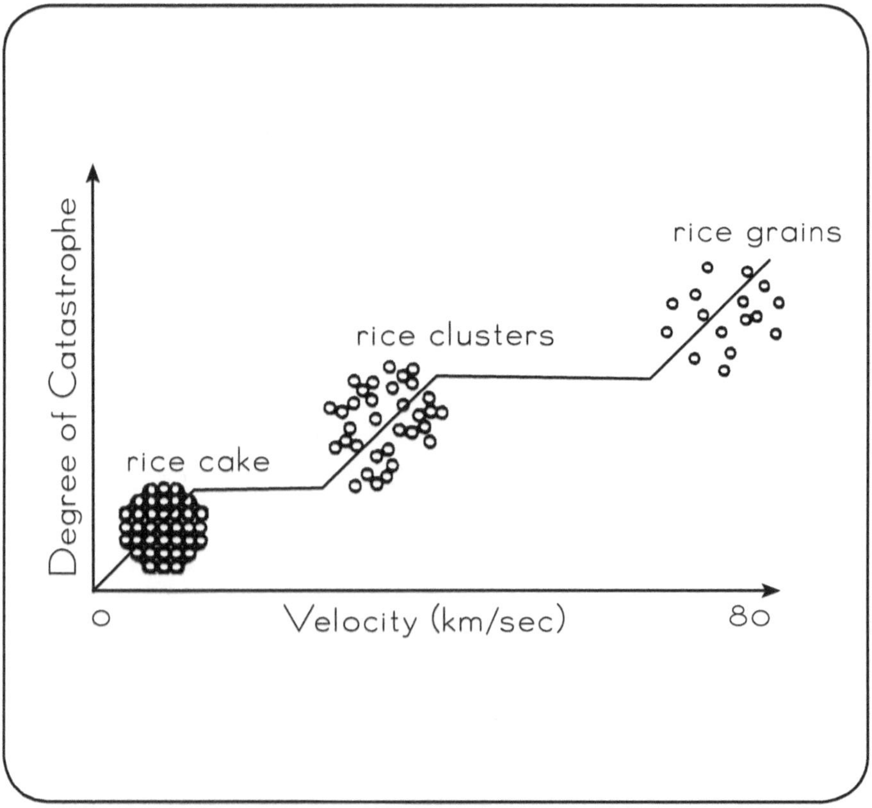

GIVEN:

- "Listen to your wife; you have two ears and one mouth—do the math" (Paul Stoltz, *The Adversity Advantage*, 2007, p. xv).

PROVE:

Represent mathematically.

GIVEN:
- "Listen to your wife; you have two ears and one mouth—do the math" (Paul Stoltz, *The Adversity Advantage*, 2007, p. xv).

PROVE:
Represent mathematically.

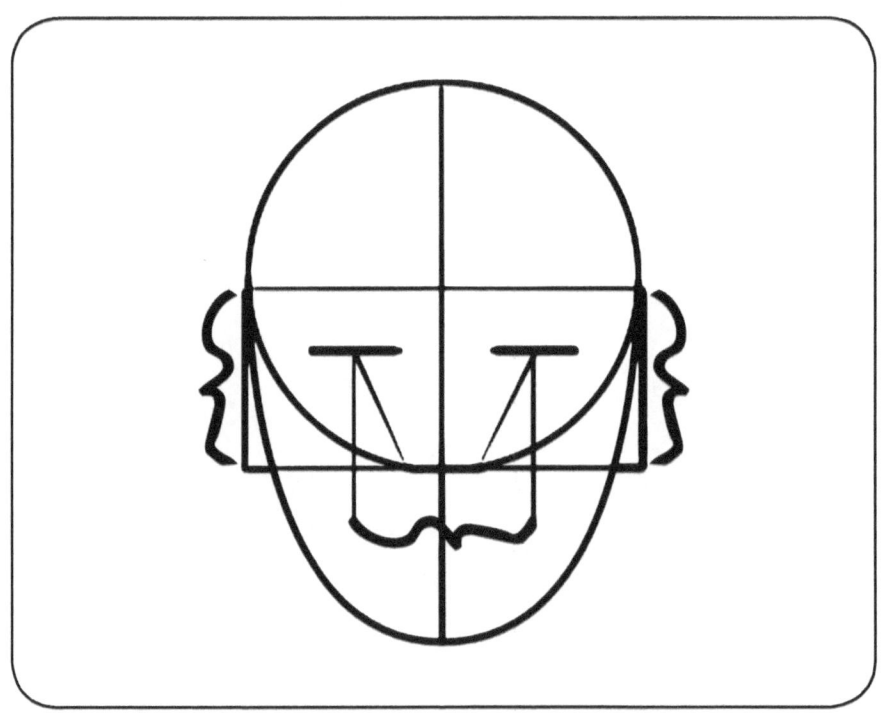

GIVEN:
- "He was a scientist. And he was killed. You do the math"
  (Boris Starling, *Visibility*, 2008, p. 158).
- The scientist was an American.

PROVE:
The culprit.

GIVEN:
- "He was a scientist. And he was killed. You do the math" (Boris Starling, *Visibility*, 2008, p. 158).
- The scientist was an American.

PROVE:
The culprit.

> The scientist was most likely killed by his own body fat.
>
> According to the World Health Organization, 64% of American adults are overweight, and half of those are morbidly obese. Only 0.005% of Americans are murdered.
>
> Death by morbid obesity is 6,400 times more likely than by murder.

GIVEN:

- "'You can't treasure me,' she said ... 'You don't know me.'
'Well, now, there we disagree. Do the math.' The last sentence
left her completely at sea. 'What math? Are we talking about
math?' 'We are now.'" (Linda Howard, *Up Close and Dangerous*,
2008, p. 250).

PROVE:
She is correct.

GIVEN:

- "'You can't treasure me,' she said . . . 'You don't know me.' 'Well, now, there we disagree. Do the math.' The last sentence left her completely at sea. 'What math? Are we talking about math?' 'We are now.'" (Linda Howard, *Up Close and Dangerous*, 2008, p. 250).

PROVE:

She is correct.

X marks the spot where treasure is buried (an intersection of longitude and latitude).

Let X = {x,y}
where -180° < x < 180° and -90° < y < 90°

Even if either x or y is known, the likelihood of striking X is infinitessimal.
